ARTILLERY

Modern Military Techniques

MODERN MILITARY TECHNIQUES
ARTILLERY

Terry Gander

Illustrations by
Tony Gibbons • Peter Sarson • Tony Bryan

Lerner Publications Company • Minneapolis

Library of Congress Cataloging-in-Publication Data

Gander, Terry.
    Artillery.

    (Modern military techniques)
    Includes index.
    Summary: Surveys modern artillery in military use,
including towed and self-propelled guns, rockets,
ammunition, and associated equipment.
        1. Ordnance—Juvenile literature. 2. Artillery—
Juvenile literature. [1. Ordance. 2. Artillery]
I. Sarson, Peter, ill. II. Bryan, Tony, ill.
III. Gibbons, Tony, ill. IV. Title. V. Series.
UF560.G36   1987   358'.1   86-10534
ISBN 0-8225-1380-3 (lib. bdg.)

Manufactured in the United States of America

        4   5   6   7   8   9   10   95   94   93   92   91   90

CONTENTS

1 Types of Modern Artillery

angles in order to hit their targets either in the side or at shallow angles. Howitzers are fired with their barrels pointing at high angles to allow their shells to fall downwards onto their targets. Both types of artillery have advantages and disadvantages, so modern weapons are usually a combination of gun and howitzer and are called gun-howitzers. Their ammunition is loaded in two parts, shell and propellant, and the barrel can be raised to almost any angle to suit the target being fired at.

Today, artillery may be either towed behind trucks or self-propelled. The self-propelled weapons move on wheels or tracks powered by their own engines and they are much more mobile

There are two main kinds of modern artillery, guns and howitzers. The ammunition for guns is loaded into them with the propellant case already fixed to the shell while howitzers have their ammunition in two pieces, the shell and the propellant charge. These are loaded separately, so the gunner can vary the charge as he chooses. The charge for a howitzer is often loaded into the breech in small bags.

Guns are usually fired with their barrels at low

The Yugoslav M56 105mm field howitzer, a typical towed field piece with split carriage trail legs, a shield to protect the crew and a muzzle brake. As well as firing high explosive (HE) shells the M56 also fires HESH to provide it with an anti-tank capability.
Specification:
Caliber 105mm; weight in action 4,532 pounds (2,060 kg); range 7.8 miles (13 km); crew 7; weight of HE shell 33 pounds (15 kg); rate of fire 16 rpm.

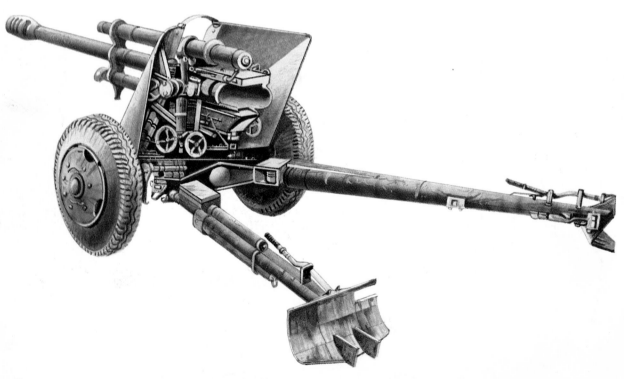

than towed weapons. Towed artillery cannot usually move across rough country easily while self-propelled artillery can. However many smaller countries still use towed artillery for it is much less costly than self-propelled artillery and does not need so much looking after.

Most modern artillery is known as field artillery. It is used to support groups of soldiers on the battlefield by making their enemies take cover, by smashing their defensive positions and equipment, and by preventing the enemy from moving across the battlefield. Long-range artillery is fired to very distant points to disrupt the lines of communication to the enemy's forward positions. This long-range fire interferes with the movement of supplies and groups of men and generally breaks up formations of tanks and other vehicles before they reach the front line. Counter-battery artillery is used to knock out the enemy's artillery before it can cause damage to friendly troops and positions.

There are some other types of modern artillery, including rockets, and they will be dealt with separately.

With the need to move large bodies of troops rapidly to any part of the world more attention is now being given to special light forms of artillery. A special section in this book mentions some of these forms of light artillery and also the special mountain guns that are very similar.

The South African 155mm G-6 self-propelled howitzer, an unusual vehicle fitted with wheels rather than tracks to enable it to operate under South African bush conditions and over great distances. The G-6 has a crew of five and can fire a 100.1 pound HE shell to a range of 18 miles (30 km), and even more using special base bleed shells that increase range. The turret can rotate only 80 degrees for normal firing and small stabilizer legs are lowered before the howitzer can be brought into action. Up to 44 projectiles can be carried in racks around the turret and more ammunition is carried inside the bush-cutting wedge at the front of the vehicle.

2
Field Artillery

Field artillery is either towed or self-propelled. It needs to be very mobile as it has to move across a battlefield at the same speed as tanks and other fighting vehicles so that it can be brought into and out of action easily. Today most armies choose self-propelled artillery as field artillery as these weapons are able to move with the tanks, leaving the towed artillery to provide a reserve and to give back-up concentrations of fire whenever the opportunity arises.

Today the most favored caliber is 155mm. The caliber of a gun or howitzer is the width of the bore down the barrel, measured from side to side. A caliber of 155mm (about 6.2 inches) enables the weapon to fire a shell sufficiently powerful to smash up large armored formations. Most nations now use weapons of this caliber or something near it — the Soviets use a caliber of 152mm. However, other weapons with both larger and smaller calibers may be found. For many years 105mm was the most popular and weapons of this caliber can still be found. Other calibers include the Soviet 122mm.

Most modern field artillery weapons can fire shells to at least 9 miles (15 km) and some can manage much more. Most modern gunners use weapons that can fire 14.4 miles (24 km) and a fortunate few have weapons that can reach out to 24 miles (40 km) with special types of shell (see

page 22). These really long-range types of weapons tend to wear out quickly and are generally kept for use against targets beyond the battlefield.

It is important that field artillery is able to fire numbers of shells in a short space of time. Most guns fire six shells a minute without difficulty, but some field guns are called upon to fire three shells in as little as 15 seconds. The object is to hit the enemy before he has time to take cover. It is difficult to reload the weapon in such a short-time, so many guns and howitzers have some form of mechanical device that loads shells and charges for them.

Field artillery must also be accurate. It must be able to fire directly at tanks should they appear in range. A 155mm shell will knock out almost any modern tank, but the gunner has to have special sighting equipment to aim the weapon accurately. Most field artillery has some form of telescopic sight for this purpose.

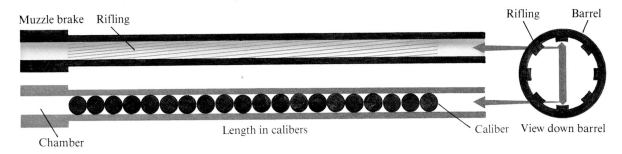

Muzzle brake   Rifling   Rifling   Barrel

                        Length in calibers   Caliber   View down barrel

Chamber

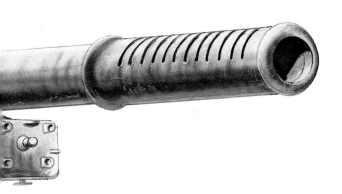

The Soviet 152mm M1937 (ML-20) gun-howitzer, a weapon that was first issued before the Second World War but is still in widespread use. This powerful weapon is towed by a two-wheeled limber on which the carriage legs rest to reduce the load on the towing tractor. The barrel is fitted with what is known as a multi-baffle muzzle brake with many side ports to divert firing gases and reduce recoil. The two large cylinders each side of the barrel are known as equilibrators and contain large springs to take some of the weight of the barrel off the layer's fire control wheels. The M1937 has a crew of nine and fires a 95.72 pounds (43.51 kg) high explosive shell to a range of 10.359 miles (17.265 km). When in action it weighs 15,994 pounds (7,270 kg) and with the limber the weight goes up to 17,765 pounds (8,075 kg) which is a heavy load to maneuver manually. The wheel tires are solid. To help balance the towed load when on the move the barrel is pulled back over the carriage, so in action the barrel would appear much longer than seen here. This weapon is still widely used in the Middle East.

9

3
Long-range Artillery

A long-range gun or howitzer can be distinguished from ordinary field artillery by its very long barrel. Generally, the longer an artillery barrel, the further the weapon will fire a shell. With long barrels, a larger propellant charge can be used. This takes more time to build up the pressure that forces the shell along the barrel. The shell leaves the barrel traveling at a much faster speed and so carries a greater distance.

When the length of the barrel on an artillery piece is stated, it actually represents the number of times the caliber can be divided into the barrel length. For instance, the British Army's FH-70 field gun barrel is 241.8 inches (6.045m) long with a caliber of 155mm. Dividing the barrel length by the caliber produces the figure 39. Thus the FH-70 is said to have a barrel 39 calibers long, written as L/39. It may also be regarded as a long-range weapon, for it has a range of 14.4 miles (24 km). Weapons with even longer barrels are in use. The South African G-5 howitzer has a barrel 45 calibers long (6.2 inches x 45 = 279 inches, or 155mm x 45 = 6.975m) and it can fire a shell to 18 miles (30 km).

Experimental weapons with barrels 52 calibers long are now being tested. Perhaps the longest barrel in use is on the American 175mm M107 self-propelled gun. The M107 barrel is 60 calibers long and it can fire a shell to a range of 19.62 miles (32.7 km).

Long barrels are difficult and expensive to make and, in addition, they have one drawback. The powerful propellant charges they fire tend to wear out the barrels more rapidly. As the barrels wear, the guns become inaccurate. It is difficult to prevent this wear, but it is possible to slow it down by reducing the rate of fire. Long-range guns may fire as few as one shot a minute, or less. They are usually reserved for special, important targets.

Long-range guns and howitzers may be either towed or self-propelled. They are best positioned close to the front lines so as to make the best possible use of the long-range. The closer the weapon is to the enemy, the further it can reach into the enemy's rear areas to disrupt supply lines and enemy movements.

Right: The FH-70 155mm howitzer can fire a nuclear shell to a range of about 10.8 miles (18 km) but it is so light it can be carried slung under a CH-47D Chinook helicopter.
Below: The Soviet S-23 180mm long-range gun

4
Counter-Battery Artillery

The Soviet 130mm M-46 field gun is used mainly for counter-battery work as it has a range of 16.29 miles (27.15 km) when firing ordinary rounds and even more when firing ammunition fitted with base bleed units. Despite being long, heavy and cumbersome it is widely used by the Red Army and many other armies around the world.

The role of counter-battery artillery is to knock out the enemy's artillery before it can attack yours. Both field and long-range artillery can be used for this job, but counter-battery artillery is specially armed for this purpose. It has the benefit of a number of devices, in the form of instruments and radar, to find out exactly where the enemy's artillery is hidden. On a battlefield artillery is very carefully concealed by camouflage and other methods, but once the weapons are fired they can be detected.

Flash produced on firing is a clue to the position of artillery, especially at night. Well-established methods of detecting artillery accurately using the flash have been developed by most armies. The sound of firing can also determine a "fix." Nowadays radar is often used. As a shell moves through the air it can be detected by radar, and computers then work out the shell's path back to the firing position.

As soon as artillery begins to fire it can be detected and counter-battery fire can be expected. This is why self-propelled artillery has a great advantage over towed artillery, for it can be built with armored protection for the crew. The crews of towed artillery have to work in the open. Self-propelled artillery can also fire and move away very quickly, while to get a towed gun or howitzer in and out of action quickly is much more difficult. Self-propelled artillery is used to fire a few shells and it is then moved away as quickly as possible to a new firing position to continue firing.

Counter-battery artillery also has to be able to move rapidly, but mainly from one target to another. When an enemy battery is located, fire must be directed against it before it moves away. To do this the counter-battery artillery units are equipped with computers to aim the weapons as quickly as possible and special methods of communication between the detection devices and the weapons. Orders have to be passed very speedily

and the weapons have to move equally quickly from one target to the next. It is here that the long-range weapons become important for they will be able to cover a far larger part of a battlefield than the shorter-ranged field artillery. Fewer of them are then needed to aim counter-battery fire over a given area.

Counter-battery artillery weapons tend to be large and cumbersome and they are usually rather heavy to provide stability and accuracy when fired. They also tend to have long barrels to provide extra range.

GIAT, a French company, produces the 155mm TR gun-howitzer. There are plans to issue this weapon to the French Army once testing is complete. An auxiliary engine is used to enable the weapon to make short moves and the range is about 14.4 miles (24 km) firing ordinary ammunition.

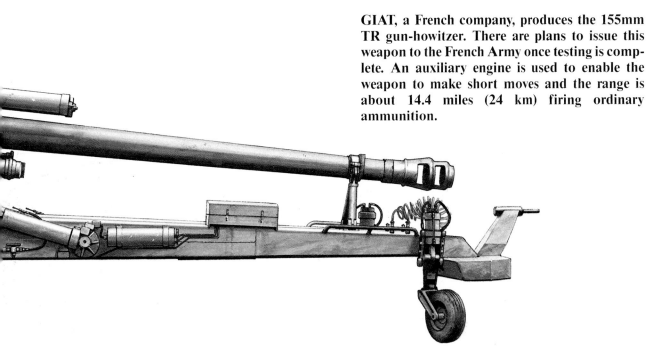

5
The Modern Towed Gun — How it Works

thick oil enclosed in cylinders and this absorbs the greater part of the recoil. On the FH-77B the rods that connect the barrel to these recoil pistons can be seen on either side of the barrel.

The FH-77B is carried on a device known as a cradle which is connected to the weapon carriage. The carriage is the main carrier for the weapon and has two legs known as trails on which the weapon rests for firing. In action these legs fold outwards; for travel they are folded together. The carriage also has wheels on which the weapon is towed, but in addition it has one very modern feature: a diesel engine that can be used to power the wheels to move the gun over short distances. Weapons such as the FH-77B are too heavy to move manually, even over a few yards, so the engine, known as an auxiliary power unit or APU, is used instead.

Other functions on the weapon are also powered

Barrel

Muzzle brake

A typical modern artillery weapon is the Swedish Bofors FH-77B, an advanced 155mm howitzer with a range of 14.4 miles (24 km). The most important part of the FH-77B, the barrel, is made of very strong steel to withstand the terrific pressures produced on firing. These firing forces push the shell from the muzzle at the end of the barrel but also force the barrel backwards to produce recoil. Recoil forces can be very violent but are absorbed by two devices on the FH-77B. One device, known as a muzzle brake, is fitted to the muzzle. This slows down the recoil slightly by directing some of the forces behind the shell to the sides of the muzzle. The main recoil force is taken up by the recoil mechanism, a series of enclosed pistons around the barrel. As the barrel moves backwards these pistons are forced through a very

by the engine, such as the raising and lowering of the barrel and the opening and closing of the heavy trail legs. Even the loading of the howitzer is helped by the engine, for the FH-77B uses a crane system to get the shells to the breech and the engine provides the power for the crane. The aimer controls all operations from his seat high on the left of the barrel. Using a control panel he not only aims the gun but works the engine and its functions. On the move, under engine power, the weight on the trail legs is taken by small extra wheels under the trail legs.

Also on the carriage are various barrel balancing devices. The barrel is very heavy so a system of springs and pistons are used to counter-balance its weight. These project on either side of the barrel and are known as equilibrators.

SPECIFICATIONS

Caliber	6.20 inches (155mm)
Length of barrel	19.95 feet (6.045m)
Weight	26,180 pounds (11,900 kg)
Length traveling	38.28 feet (11.6m)
Width traveling	8.75 feet (2.65m)
Height traveling	9.24 feet (2.8m)
Elevation/depression	+70°/−3°
Traverse	60°
Rate of fire	3 rounds in 12 seconds
Maximum range (normal shells/ specific shells)	14.4 miles (24,000m)/ 18 miles (30,000m)
Speed (towed/APU power)	42 mph (70 km/h)/ 4.8 mph (8 km/h)
Shell weight (HE)	94.38 pounds (42.9 kg)

Equilibrator

Aimer's position

Recoil mechanism

Barrel clamp

Drive wheel

Trail leg

Trail wheel

APU

6
The Modern Self-Propelled Weapon — How It Works

Most modern self-propelled artillery weapons are now fully protected, the weapon and crew enclosed in an armored turret. However some weapons are simply mounted on a tracked vehicle. One of these is the American M110 self-propelled artillery howitzer which has a caliber of 8 inches/203mm and can fire an atomic shell.

The M110 barrel uses the power from the main vehicle engine to power everything. The raising and lowering (elevating and depressing) of the barrel is powered, as is part of the loading process, including the pushing of the shell into the breech (ramming). The balancing of the heavy barrel on the tracked carriage is carried out by equilibrators in exactly the same way as on a towed weapon, and

they can be seen on either side of the barrel on the M110. Before firing, a large powered spade is driven into the ground at the rear of the M110 to absorb some of the recoil forces.

On an enclosed self-propelled 155mm howitzer such as the American M109 very little can be seen of the weapon other than the barrel poking through the vehicle turret. The turret is armored to protect the crew who do not normally have to leave the M109 once in action, for the ammunition is stacked inside in special racks. As on the M110, everything is powered by the vehicle engine, but on the M109 the turret can rotate (traverse) through a full circle for firing. This has many advantages in action when fire might have to be switched rapidly from one target to another. On the M110 the barrel has only a slight degree of movement before the entire vehicle has to be moved round to the new direction.

Self-propelled artillery vehicles usually have a crew of four or five men. One man drives the vehicle, with the others loading and aiming the weapon. One soldier is always in charge, taking orders over a radio for aiming the barrel. Sometimes the crew themselves decide what their target will be and use an on-board computer to work out all the necessary aiming data. The same computer

The M110A2 8-inch/203mm self-propelled howitzer can be used to fire nuclear shells to a range of 17.46 miles (29.1 km), using a special rocket-assisted projectile (RAP). The M110A2 has a crew of five and uses a small crane to lift projectiles from ground level up to the breech.

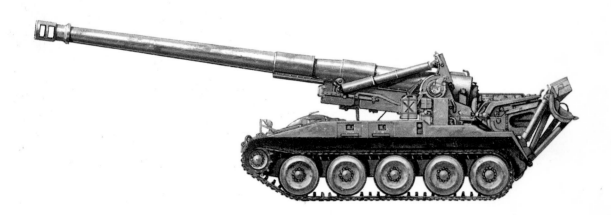

US M110 short-barrel 203mm Howitzer

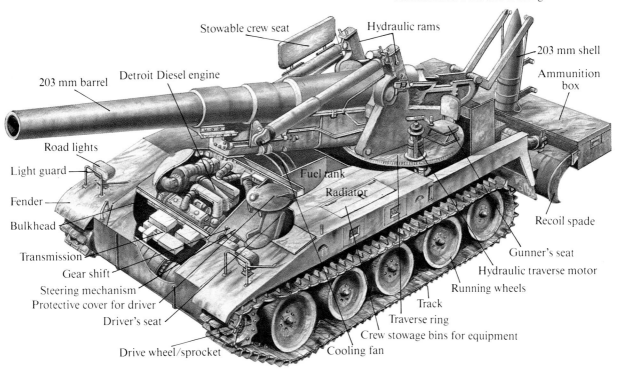

Stowable crew seat

Hydraulic rams

Ammunition hoist and loading mechanism

203 mm shell

203 mm barrel

Detroit Diesel engine

Ammunition box

Road lights

Light guard

Fuel tank

Fender

Radiator

Bulkhead

Recoil spade

Transmission

Gear shift

Gunner's seat

Steering mechanism

Hydraulic traverse motor

Protective cover for driver

Running wheels

Driver's seat

Track

Traverse ring

Drive wheel/sprocket

Crew stowage bins for equipment

Cooling fan

is used to find out other facts, such as the exact location of the M109 at any one time, the stock of ammunition and fuel, and for storing lists of targets ready for instant use. Radios are carried to enable the gunners to talk to other vehicles and the local headquarters.

The American 155mm M109A2 self-propelled howitzer is one of the most widely used weapons of its type in the Western world. Despite its weight and bulk the aluminum armor used for protection makes the vehicle light enough to "swim" when crossing water obstacles.

7
Organization of an Artillery Group

not necessarily require the weapons themselves to be rushed from one point to another. In many instances the long range of modern artillery enables it to remain in one spot, simply diverting attention from one firing task to another.

To do this the various weapons have to be organized into groups on the battlefield, working together. They are not positioned close to one another but rather spread out to avoid detection and joined together by radio. To enable these groups to be controlled by one central point it has been found that their size must be limited. Experience has shown that the optimum size is about three batteries.

In these three-battery groups, each battery has between six and eight weapons. At times the three

Modern artillery has to be able to operate with great flexibility on a battlefield. At one point in a battle it may have to be massed together to let loose a heavy bombardment of fire in order to support an attack, and a few moments later it may have to disperse to fire at a series of local targets. Then, just as quickly, it might have to join up again to provide fire to support an advance which is trying to get through a hold-up caused by an enemy strongpoint. This concentration and dispersal of fire does

Right: The M992 Field Artillery Ammunition Support Vehicle (FAASV) uses a much-modified M109A2 self-propelled howitzer hull to carry ammunition to batteries in front-line areas. It has a crew of two.

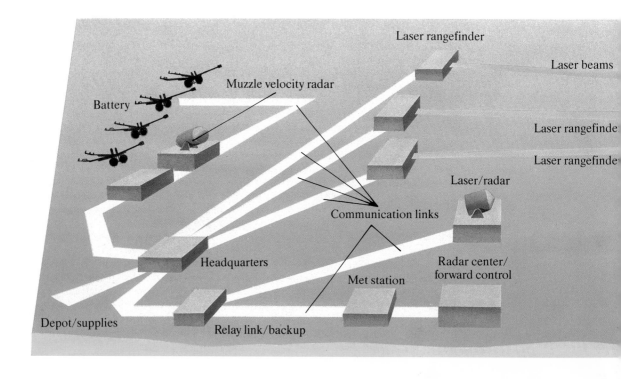

A typical layout of a battery fire control system with laser rangefinders being used to determine target ranges accurately. This data is sent to the battery fire control post where it is modified by other inputs such as that from met sensors, radar data and muzzle velocity radars before accurate fire control orders are transmitted to the guns in the battery. Links can be either by telephone wire or by radio.

batteries will be used as separate units, firing at different targets, while on other occasions they will all fire together at the same target. These groups are known by various names. In some armies they are battalions, in others artillery groups, and in some (including the British Army) they are known as regiments.

Numbers of groups are joined together for special fire missions, such as a massed artillery bombardment of a special target, but normally they operate as one unit. They have their own central headquarters unit to issue orders to the three batteries, and this same central headquarters is usually responsible for ordering up fresh ammunition and supplies once the group is in action. The headquarters also controls the important radio network that links every battery and weapon together. This is especially important with self-propelled artillery units, where individual weapons often operate without even seeing other vehicles from the same battery. In towed artillery batteries, the weapons are controlled from more central points, for they rarely operate by themselves.

8 A Battery in Action

Both towed and self-propelled artillery is organized into six- or eight-gun batteries. Each battery consists of a headquarters unit and two sections, each with three or four weapons. Sometimes these may be further organized into sub-sections but whenever possible the sections will operate as a battery. This is to make sure that the full power of the artillery can be employed to get as many shells on a target as quickly as possible. The headquarters unit is responsible for giving orders to the weapons, deciding at which targets they will aim and when they will fire. It also organizes fresh ammunition, fuel, and other supplies for the individual weapons and crew members.

In action, the sections usually operate as a sub-unit. They can fight for periods without having to take constant orders from the headquarters unit on every fire mission, but they are in touch with the headquarters by radio at all times. This applies particularly to towed artillery because each weapon does not necessarily have a radio. They may be connected to a local section control post by telephone line or even by simply shouting from one position to another.

On each weapon, towed or self-propelled, one man is in charge. A 155mm self-propelled howitzer has a crew of four or five men. A towed 155mm howitzer may have a crew of seven men or more, as there is more labor involved in moving the weapon around. In every crew one man will actually aim the weapon and operate the controls that move the barrel around — this is known as "laying" the weapon. The rest of the crew will be involved with the ammunition. One man will unpack it from the crate, box, or pallet in which it is supplied to the weapon position (on a self-propelled weapon the ammunition is usually already on the vehicle). Another man will check the ammunition, set the fuze or prepare it for use, while another will take the shell to the weapon for loading into the gun. On some towed weapons the shell has to be rammed into the breech by hand; on other weapons this operation is powered. The aimer usually fires the weapon by pulling a lever or pressing a foot pedal.

Loading any modern artillery weapon in action is hard work. A 155mm shell weighs about 95 pounds (43 kg) and it often has to be carried by hand during some part of the loading sequence. A 155mm battery can easily fire 20 tons of ammunition in 20 minutes and that weight has to be moved by hand quite a lot. Even with modern loading and handling methods gunners have to be strong men.

A battery of American 8-inch/203mm M115 towed howitzers in action under typical front line conditions of dust, noise and smoke. The howitzers are placed rather more closely than they would be under wartime conditions. The howitzers are the same as those used on the M110 self-propelled howitzer.

9
Ammunition

HE

Illuminating

Smoke

In defense terms the projectile or artillery shell is the gunner's weapon, for it is the shell that actually destroys the enemy or his equipment and position. The gun or howitzer is simply the delivery system. To a gunner, the shell is important, and the main type he uses is known as high explosive, or HE.

The HE shell consists of a mass of explosive packed into a steel body. When it is fired the grooves within the gun or howitzer barrel, known as the "rifling," twist the shell around its length to ensure that it flies along its intended path. As it reaches the target the shell may be exploded by means of a fuse.

There are several types of fuse. One shell has to hit something to activate its fuse. In another, a small clock times the length of flight, exploding the shell after a certain time. A special type of fuse, known as a proximity fuse explodes the shell when it is at a certain height above the ground. These shells are usually fired from howitzers, for they have to fall to the ground at steep angles to operate the proximity fuse. When the shell goes off it produces a violent blast, bursting the steel shell casing to send steel splinters flying in all directions. By exploding the shell above the ground the splinters cover a wider area.

Not all shells are HE. Some produce smoke clouds which are used to hide movements from the enemy or to indicate targets to attacking aircraft. Parachutes are unfolded from the rear of other shells and these are used to suspend bright lights over a battlefield at night in order to illuminate it. Other cargo shells carry special loads, such as small mines, that are scattered in the path of tanks. There are many different types of special loads contained in cargo shells.

To increase their range, special shells have been developed. Some rely on a carefully designed outline to fly further, while others use rockets in their tail to push the shell along. A base bleed (BB) version burns a substance in its tail. This burning breaks up the air currents behind the shell that tend to slow it down so the shell can travel further. A recently developed shell uses laser detectors in its nose to guide it along laser beams directed from the ground to a target, and uses small wings on the shell body for the final stages of guidance.

A graphic indication of how enhanced range ammunition can improve the overall range of projectiles, although it must be stressed that rocket assisted projectiles (RAPs) can be very inaccurate

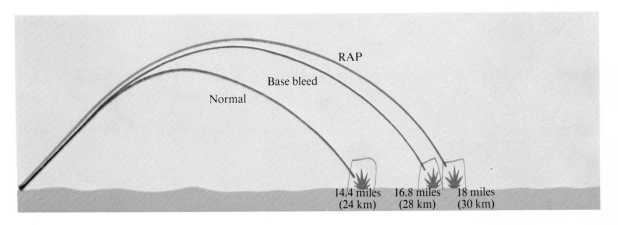

RAP

Base bleed

Normal

14.4 miles (24 km)   16.8 miles (28 km)   18 miles (30 km)

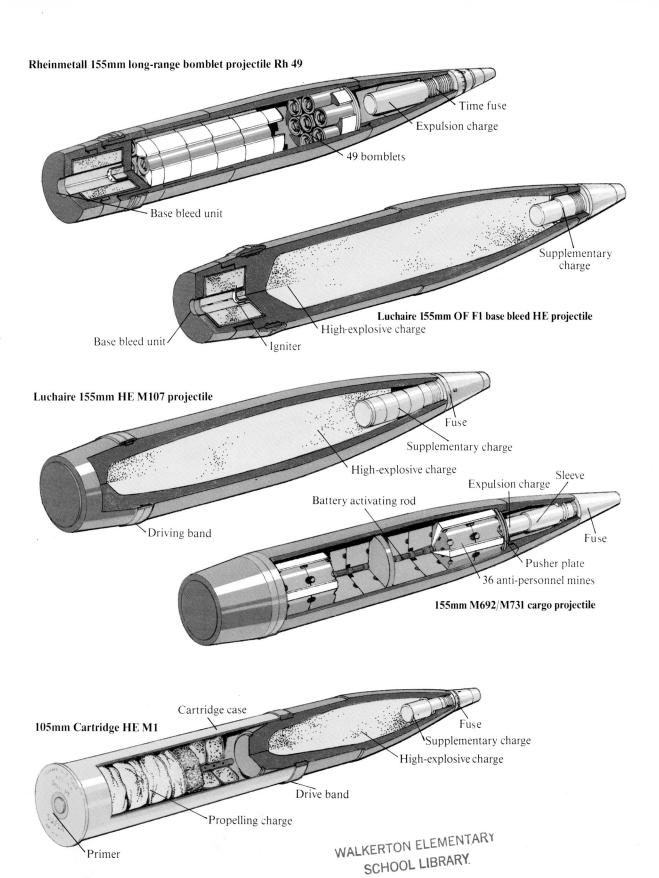

Rheinmetall 155mm long-range bomblet projectile Rh 49

Time fuse

Expulsion charge

49 bomblets

Base bleed unit

Supplementary charge

Luchaire 155mm OF F1 base bleed HE projectile

Base bleed unit

Igniter

High-explosive charge

Luchaire 155mm HE M107 projectile

Fuse

Supplementary charge

High-explosive charge

Driving band

Sleeve

Expulsion charge

Fuse

Battery activating rod

Pusher plate

36 anti-personnel mines

155mm M692/M731 cargo projectile

Cartridge case

Fuse

105mm Cartridge HE M1

Supplementary charge

High-explosive charge

Drive band

Propelling charge

Primer

10 Finding the Target — the Observation Post

The Israeli David fire-control computer can handle fire-control data for up to six (or more) guns and its memory can store data for up to 28 targets.

Gunners rarely see the targets at which they are aiming, so to find out exactly where their shells are landing they use a forward observer. This trained gunner takes up a position well forward in the front-line close to the targets. He selects the targets and keeps the battery informed as to where their shells are landing. Messages are sent back by radio which enable the gunners to make corrections to their aim when necessary. The forward observer often operates from an armored vehicle traveling with tanks.

The observer is equipped with a number of instruments which help him make his aiming corrections. One is a rangefinder, usually laser-powered, with which he can tell how far a target is from his own position. With this information he can then work out or use a small computer to calculate the distance of the target from the battery. Additional information comes from special surveying instruments and simple binoculars. Computers are being used increasingly for forward field artillery observations for the old method of firing a single shot to determine how far out the first shot and then making gradual change to the aim until the shells are on target is no longer effective. The first shells have to be on target now for otherwise the enemy will either take cover or simply move their vehicles away from the target area.

As with so much of modern artillery, good communication is the key to this operation. The forward observer may use radio to talk to his battery but radio messages can be overheard by an enemy so telephone lines are used whenever possible. When radio messages have to be sent, a device electronically compresses a lengthy word message into a very short space of time. At the receiver radio set another electronic device "unscrambles" the message back to normal length. This prevents an enemy overhearing the message and using their radio instruments to determine the position of the observer.

Very often, forward observers operate right in the front lines. They may even be behind enemy lines in specially built hiding places in order to observe long-range artillery or counter-battery fire.

Forward observers operate in small groups of two or three. They usually stay away from their batteries for long periods and thus have to look after and defend themselves at all times.

Right: A British Army forward observation officer equipped with a night-sight over which is positioned a laser rangefinder and target indicator. The indicator fires bursts of laser energy to indicate targets to attack aircraft.

The **RATAC** battlefield radar is used by the US Army and by the armies of France, West Germany and several other nations. It is used to detect the position of tanks, troops and low-flying aircraft and can be used for artillery fire direction and control. It can detect groups of men at up to 4.8 miles (8 km) and 9 miles (15 km) against tanks.

11 American Artillery

American front-line artillery is now almost entirely self-propelled. The largest weapon in service is the 8 inch/203mm M110 howitzer, which can fire an atomic shell. The most numerous is the 155mm M109 howitzer. The M109, used by many countries, can fire a wide range of different types of shells from HE to various forms of poison gas. It is a large vehicle but relatively light as much of its armor protection is made from aluminum. In fact it is so light that it can float across rivers driving itself with its own tracks.

The most recent M109 has a range of 10.86 miles (18.1 km) and the HE shell weighs just under 94.6 pounds (43 kg). The barrel is 39 calibers long, though to produce longer ranges some nations are putting new barrels into their M109s, some of them 45 calibers long. A machine gun is carried on the roof for anti-aircraft and other defense. Its crew numbers six. Many nations use the M109, and one, Israel, has used them in action many times.

Not all American artillery is self-propelled, for some special units, such as the US Marine Corps, carry out operations where their equipment has to be light enough to be carried in transport aircraft or even slung under helicopters. For their operations towed artillery is preferred. They use a towed 155mm howitzer known as the M198, a very simple light howitzer that is unusual in not having an auxiliary engine to power the wheels and barrel movements. Everything has to be moved by a towing vehicle or by the gunners themselves. The M198 fires the same ammunition as the self-propelled M109 but a new propellant charge system gives the shells a range of up to 14.4 miles (24 km).

The American forces also include airborne units that parachute into action and so need small, light artillery. At present the US Army airborne units use a small towed howitzer with a caliber of 105mm, known as the M102. This fires an HE shell weighing 46.2 pounds (21 kg), but the range is only 6.9 miles (11.5 km). This is because the weight has to be kept as low as possible, so the barrel is only 32 calibers long.

Many American gunners now want a weapon with a longer-range and are interested in a British 105mm gun, known simply as "the Light Gun." This gun performed so well during the Falkland Islands war that many nations are now buying it. The US Army is expected to replace their M102s with a number of these. The Light Gun has a range of 10.32 miles (17.2 km) and fires a shell weighing 35.4 pounds (16.1 kg), but it is slightly heavier than the M102.

The American M198 155mm howitzer, showing its wide arc of elevation (left) and the extent the barrel can be traversed each side of a center line; for wider traversing the entire gun can be quickly moved, through 180 degrees if required.

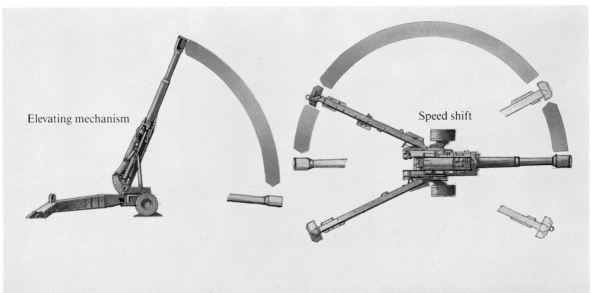

Elevating mechanism

Speed shift

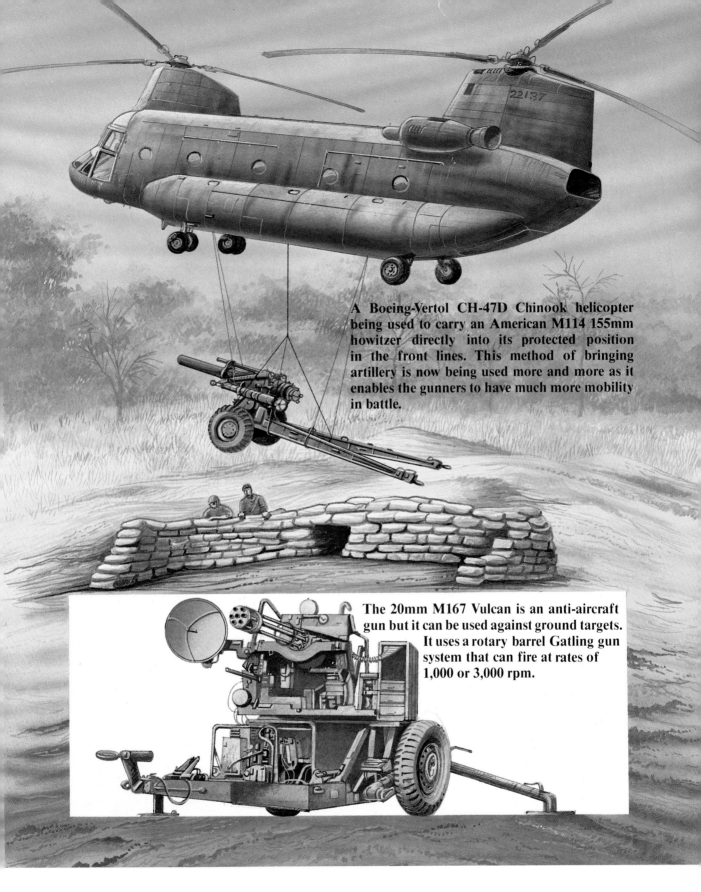

A Boeing-Vertol **CH-47D Chinook** helicopter being used to carry an American **M114 155mm** howitzer directly into its protected position in the front lines. This method of bringing artillery is now being used more and more as it enables the gunners to have much more mobility in battle.

The 20mm **M167 Vulcan** is an anti-aircraft gun but it can be used against ground targets. It uses a rotary barrel Gatling gun system that can fire at rates of 1,000 or 3,000 rpm.

12 Soviet Artillery

The Soviet Army has more artillery than any other nation and uses it in great numbers in action. It uses a wide range of different types of weapons, many of them towed. Only a relatively few front-line armored formations use self-propelled artillery.

Soviet weapons are mainly in two calibers, 122mm and 152mm. This greatly eases the problem of ammunition supply to the front line. The Soviet Army uses only a few types of weapon with their main frontline formations. Two of these are self-propelled howitzers, the 122mm 2S1 and the 152mm 2S3 (these names are Soviet code words). A third, known as the 152mm 2S5, about which little is yet known, is a gun that can fire an atomic shell. There is also a Soviet self-propelled 203mm howitzer, but not even a photograph has yet been seen of that. The 122mm 2S1 and 152mm 2S3 are very similar in layout to the American M109; in fact the 152mm 2S3 is almost identical to its American equivalent. Even their shell weights and ranges are similar.

There are many types of Soviet towed artillery weapons, many with calibers of 122mm and 152mm, some of them rather old designs — even dating back to before the Second World War. Compared with many modern Western designs Soviet artillery weapons are simple and basic, but they are very strong weapons that need only a minimum of care and maintenance. A typical Soviet weapon is the 152mm D-20 which fires a 95.7 pound (43.5 kg) shell to a range of 10.44 miles (17.4 km). Many Soviet batteries use a lighter weapon, the 122mm D-30 howitzer which rests on a three-legged carriage when in action. This enables the barrel to be swung very quickly through an angle of 360 degrees. This howitzer is meant to be used to destroy any tanks that come within range, a combat feature shared by most

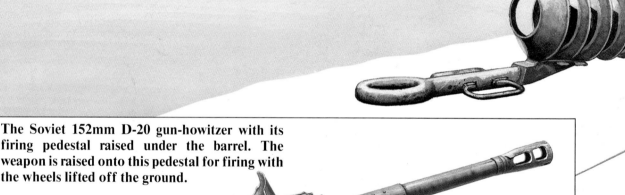

The Soviet 152mm D-20 gun-howitzer with its firing pedestal raised under the barrel. The weapon is raised onto this pedestal for firing with the wheels lifted off the ground.

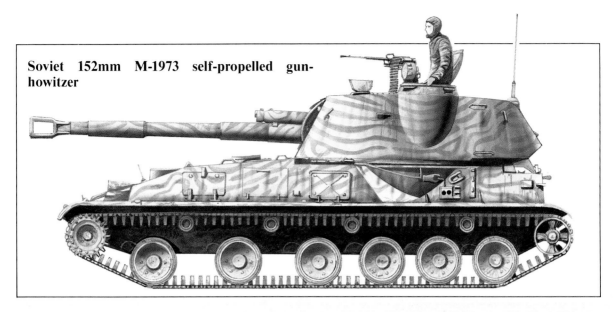

Soviet 152mm M-1973 self-propelled gun-howitzer

Soviet artillery weapons. The D-30 fires a 47.74 pound (21.7 kg) HE shell to a range of 9.24 miles (15.4 km) but a special rocket-assisted shell (rocket assisted projectile, or RAP) can reach nearly 13.2 miles (22 km).

For special counter-battery use the Soviet Army has two long-range weapons, the 180mm S-23 gun and the 130mm M-46 gun. Both are powerful guns, especially the 180mm S-23 which has a maximum range using an RAP of 26.28 miles (43.8 km), while the 130mm M-46 can reach 16.29 miles (27.15 km) with a normal HE shell.

The 122mm D-30 howitzer, one of the most widely-used of all current Soviet artillery weapons. This weapon can fire a 47.87 pounds (21.76 kg) high-explosive shell to a range of 9.24 miles (15.4 km) and the carriage can traverse through 360 degrees on its tripod carriage legs.

13 Rocket Artillery

Modern armies use rockets alongside their guns and howitzers because rockets have particular uses on the battlefield. Whereas artillery weapons provide a fairly accurate method of placing shells on selected targets, rockets are relatively inaccurate but are used to cover a wide target area. Rockets are known as area weapons. Used in large numbers, they can create a great deal of damage to an enemy force. Rockets are cheaper and easier than artillery weapons to make and to use, and they have one other advantage over a gun or howitzer. When used in heavy salvos they create a dreadful noise that strikes fear into even the most battle-hardened soldier, and so they are a morale-sapping weapon of considerable power.

Most artillery rockets are simple devices fired from launcher tubes or racks carried on trucks. Only a few towed launchers may be found. Most rockets have relatively short ranges compared with guns or howitzers. For instance, a Soviet 140mm rocket only has a range of about 6 miles (10 km), but it is fired from truck-carried launchers that fire up to 16 rockets at a time. These launchers are not used singly or in pairs but in batteries, all firing together. They can cover a large area with masses of high explosive in a very short time. Anything in the target area will be either destroyed or left in a state which makes further combat almost impossible.

The Soviet armed forces make great use of multiple rocket launchers, some with calibers as large as 250mm. Most have relatively short ranges and rely on massed firing for their effect. Some Western nations still use artillery rocket launchers, one being Spain which has an artillery rocket system known as Teruel.

A new type of artillery rocket is now being used. This fires a long-range rocket that is used to break up enemy armored formations before they have a chance to enter a battle. The rockets carry small anti-tank mines and are used in large numbers. Typical of these is the American Multiple Launch Rocket System or MLRS which fires a 227mm rocket to a range of over 18 miles (30 km). Tracked vehicles carry up to six of these rockets which are fired in ripples after which the launchers are rapidly reloaded.

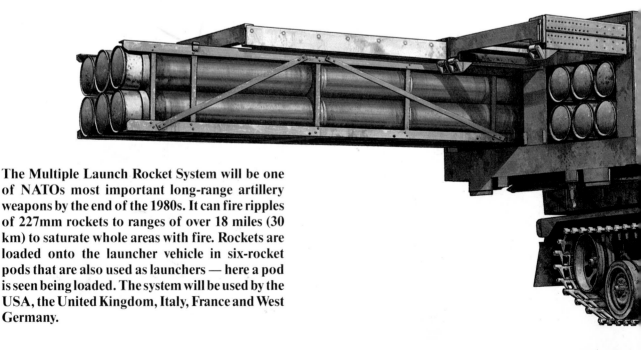

The Multiple Launch Rocket System will be one of NATOs most important long-range artillery weapons by the end of the 1980s. It can fire ripples of 227mm rockets to ranges of over 18 miles (30 km) to saturate whole areas with fire. Rockets are loaded onto the launcher vehicle in six-rocket pods that are also used as launchers — here a pod is seen being loaded. The system will be used by the USA, the United Kingdom, Italy, France and West Germany.

A battery of Czech 130mm M51 32-round rocket launchers firing their 53.24 pounds (24.2 kg) rockets in ripples from their box-like launch frames. These launchers are mounted on Praga 6 x 6 trucks and use spin to stabilize the rockets in flight.

The Soviet 132mm BM-13-16 Katyuscha was used by the Red Army during the Second World War and is still used by some nations. It can fire salvos of rockets to a range of about 5.4 miles (9 km) but it is not very accurate.

14
Long-Range Artillery Rockets

and are among the most powerful of all artillery weapons.

The average long-range artillery rocket is large and may weigh up to 5,500 pounds (2,500 kg), or more. A typical example is the American Lance which has a length of 20.36 feet (6.17m) and a body diameter of 22.4 inches (560mm). The range of Lance may be anything up to 72 miles (120 km). The Soviet counterpart to Lance is known as FROG (Free-Flight Rocket Over Ground) of which there are several types. One of the latest is the FROG-7 with a length of 19.8 feet (6m) and weighing 5,500 pounds (2,500 kg). It has a range of 42 miles (70 km).

Both of these large rockets have one thing in common. They are launched and then left to travel

The long-range artillery rocket is a special battle-field weapon used to deliver heavy explosive warheads deep into the enemy's rear areas. Some of these long-range rockets carry atomic warheads

The American Lance nuclear missile has all its flight data fed into it by a computer before launch and thereafter travels to its target without any other controls. It is light enough to be carried by helicopter.

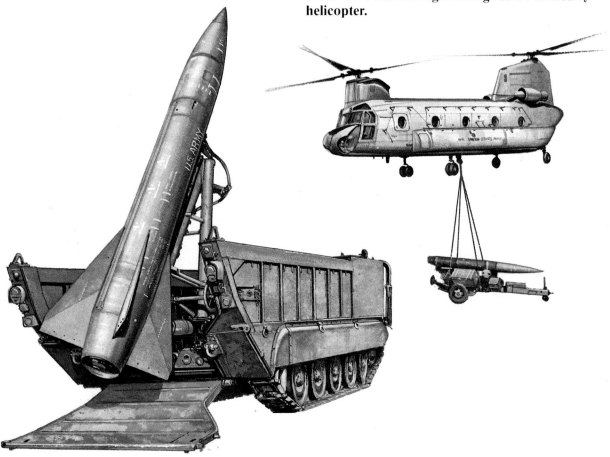

towards their targets without any guidance once they are in flight. All the target information is fed into the rocket and launcher before firing so once in flight their flight paths cannot be altered or interfered with. This is different from the much larger intercontinental missiles that are usually guided throughout their flights from the ground.

Long-range artillery rockets are transported to the battlefield on tracked or wheeled launchers. Their launchers carry only one rocket which is raised to the launching angle by a ramp or rail. The fuel used to propel the rocket may be liquid or solid and is usually loaded into the rocket well before it is carried into action. A computer on the launch vehicle is used to insert target information into the rocket's own internal guidance system where it is stored ready for use after launch.

Long-range artillery rockets are such powerful weapons that they are controlled in action by very high command levels and not fired unless a high-ranking commander gives permission. Targets are carefully selected, usually built-up areas through which enemy forces will have to pass to the battle zone or areas which are being used for storage of supplies or equipment. Ports or large railway centers are likely targets, as are large military training areas.

A Soviet FROG-7 battlefield support missile mounted on its ZIL-135 launch vehicle. The rocket weighs 5,500 pounds (2,500 kg) and has a range of 42 miles (70 km). It can be fitted with a nuclear warhead.

15
Anti-Tank Guns

Anti-tank guns are towed guns that fire solid projectiles to punch holes through enemy tanks or other armored vehicles. To do this anti-tank guns have very long barrels and use special, powerful charges to propel the solid metal projectiles at very high speeds, known as muzzle velocities. The muzzle velocity denotes the rate at which the projectile leaves the gun muzzle. The ammunition used is known as "shot," as it is completely solid and does not have any HE content.

As their name implies, anti-tank guns are a very specialized form of gun, and today not many remain in service. Their place has been taken by anti-tank guided missiles that use a variety of methods for burning through tank armor using hollow charge warheads. The anti-tank gun relies upon the shot speed and weight to simply punch its

way through the armor, and the guided missile or close-range weapon using hollow charge warhead is now considered to be more efficient at penetrating armored targets. But the anti-tank gun is still used by many nations, for it is an accurate method of knocking-out enemy tanks and it can be used at quite long ranges of up to 1.2 miles (2 km). Many infantry anti-tank weapons can only be used at ranges as close as 165 feet (50m).

Most modern artillery can be used to knock out enemy tanks by shell power alone. Any 155mm HE shell will usually destroy a tank by sheer weight of explosive, but many howitzers and guns can fire a shell known as High Explosive Anti-Tank, or HEAT, which has a warhead that produces a jet of very hot flame that burns its way forward through armor plate. HESH is a special type of anti-armor projectile on which the warhead spreads out as it hits tank armor and is then detonated. The force of the explosion against the armor creates such a shock that large pieces fly off internally, knocking out the tank and the crew. HEAT and HESH rounds can be fired from most artillery weapons, making them into tank killers. But HEAT rounds have to be fired at close ranges of about 1.2 miles (2 km), and the weapon has to be equipped with special sights.

Thus specialized anti-tank guns are now being removed from service with many armies. For all their power against armor, they are expensive, for the propellant charges and the long barrels together mean that the gun wears out and has to be replaced more often than normal weapons.

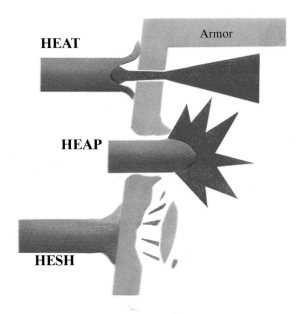

HEAT

Armor

HEAP

HESH

Left: Three of the main types of anti-tank projectiles used today. HEAT burns its way through armor by producing a jet of intensely hot flame. HEAP punches its way through by force of weight and energy and then explodes inside the target, while HESH squashes itself flat against the target before exploding to produce "scabs" of metal inside the target.
Right center: An anti-tank gun fires its projectiles over very flat trajectories but it must be able to traverse quickly over a wide arc to follow tank targets crossing its front.
Right bottom: The Soviet 85mm SD-44 gun can be used as a field gun or an anti-tank gun and has a small engine on the carriage to propel it across a battlefield.

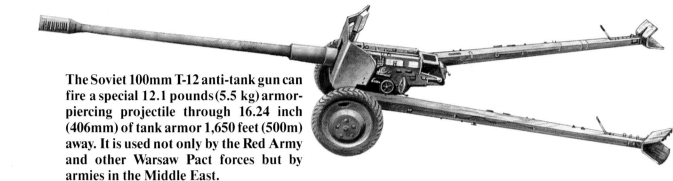

The Soviet 100mm T-12 anti-tank gun can fire a special 12.1 pounds (5.5 kg) armor-piercing projectile through 16.24 inch (406mm) of tank armor 1,650 feet (500m) away. It is used not only by the Red Army and other Warsaw Pact forces but by armies in the Middle East.

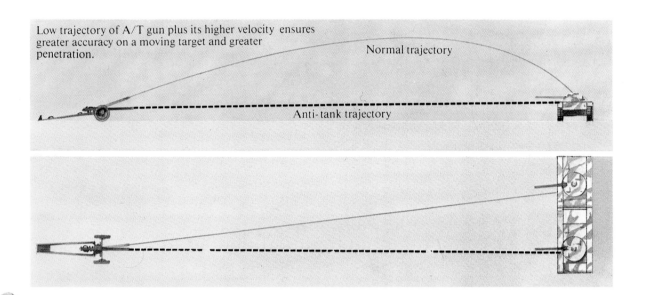

Low trajectory of A/T gun plus its higher velocity ensures greater accuracy on a moving target and greater penetration.

Normal trajectory

Anti-tank trajectory

16 Transport and Tractors

Specialized vehicles are now used to bring towed artillery into action. Most of these vehicles resemble ordinary trucks but have more powerful engines than is normal and carry special equipment for their artillery-towing role. Ordinary trucks can be used to tow artillery, but most armies consider the towing role so important that they prefer to use special vehicles.

Most artillery tractors are now wheeled, although some forces still use special tracked vehicles. In general, tracked tractors are slower but they do have the advantage of being able to pull heavier loads. A wheeled tractor is a large vehicle for not only does it have to bear the load of the towed weapon, but the gun crew, a load of ammunition and other equipment for the gun. A typical modern artillery tractor has either a large cab for the driver and crew or some form of cab on the back of the vehicle. The crew of a large howitzer may be eight men or more, and wherever possible they are carried on the tractor, together with all their personal kit and equipment. As much ammunition as possible is carried, either in racks on the tractor or in prepacked pallets that are lowered to the ground close to the weapon using a small crane on the tractor. Space on the tractor has to be found for the weapon's sighting equipment (which, for safety, is not normally carried on the weapon when traveling), various spares for the weapon, and items such as towing cables, tools and camouflage nets and poles. Many tractors carry a powerful winch and cable for dragging the weapon out of pits or soft ground.

The artillery tractor is more than just a truck. Most trucks do not have the ability to cross rough ground, but artillery tractors often have to do this. Tracked vehicles come into their own when the terrain is rough. However, as they tend to be slower than wheeled tractors, much more expensive, and take a lot more looking after than wheeled tractors, they are not used as much.

Specification:
Foden Medium Mobility Gun Tractor
Seating 1 + 10; Weight loaded 60,368 pounds (27,440 kg); Length 30.228 feet (9.16 m); Width 8.25 (2.5 m); Height 12.375 feet (3.75 m); Engine Rolls-Royce 6-cylinder diesel developing 305 hp; Maximum road speed 48 miles per hour (80 km/h); Maximum towed load 20,460 pounds (9,300 kg)

Above: A South African Defense Forces SAMIL 100 tractor towing a 155mm G-5 howitzer. This tractor carries the howitzer's eight-man crew and ammunition stacked in pallets behind the main crew cab. It has provision for an anti-aircraft machine gun on the roof.

Below: A British Army Foden Medium Mobility 6 x 6 tractor towing a FH-70 155mm field howitzer. This tractor carries the FH-70's crew in the shelter behind the driver's cab and a small hydraulic crane is used to lower pallets of ammunition from the storage area behind the cab. The driver's and gun crew's cabs are heated and have space for the crew's personal kit and weapons while there is a roof hatch over the driver's cab for a machine gun. The tractor can wade through water three feet (1m) deep and on the road can travel at speeds of over 48 miles per hour (80 km/h).

17 Airborne Artillery

neuvered by hand once on the ground, for it is unlikely that anything larger than a Jeep could be para-dropped to tow weapons.

Airborne artillery first appeared during the Second World War and was a special type known as recoilless artillery. The idea was that as the shell was fired forward from the barrel, the recoil would be counteracted by an equal weight of dense gas fired from an open breech to the rear. With equal blasts being produced at both ends of the gun barrel, there should be no recoil. This allowed a much lighter barrel and carriage to be made and

The airborne soldier arriving in battle by parachute is now an established part of modern warfare. But the parachute-carried soldier has to have weapons to support him in action, and this includes artillery. A special form of artillery has been developed that can be para-dropped onto the battlefield. Such weapons need to be small and light, not an easy matter for artillery. The guns have to be sturdy enough to absorb the shock involved in landing by parachute, yet light enough to be ma-

The principle of the recoilless gun is that when the weapon is fired to propel the projectile from the muzzle, a mass of gas is ejected to the rear through a cone-shaped aperture. By careful design the two forces cancel each other out and no recoil results. Below is a Yugoslav 82mm M60 recoilless gun clearly showing the breech device through which the firing gases can escape. They tend to produce a cloud of dust and debris behind the gun position.

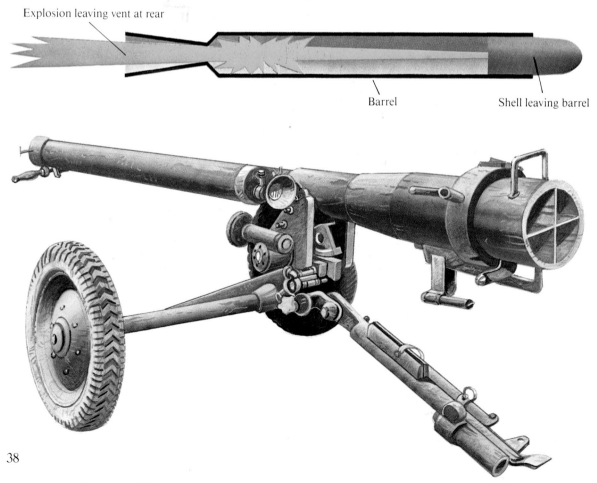

Explosion leaving vent at rear

Barrel

Shell leaving barrel

seemed to be just the thing for airborne warfare and para-dropping. The idea worked and the gun was dropped into action, but it was soon discovered that the back blast produced large clouds of dust and smoke revealing the gun position. The back-blast was also dangerous to anyone within 165 feet (50m) to the rear of the breech. In addition, the flash and noise produced on firing was so great that special ear protection was needed for the crews. Recoilless guns have their limitations, but they continue to be used, as their light weight makes them ideal for para-dropping and the drawbacks have to be tolerated.

Some conventional artillery is still used for the airborne role. The American 75mm M116 light howitzer was designed to be airborne, and may still be found in use by some of the smaller nations, but these days the tendency is to land artillery from helicopters or landed transport aircraft. This ensures that it lands in one piece and ready for immediate action, for para-dropped weapons have to be unpacked and prepared before they can be fired. The Americans use the 105mm M102 howitzer for their airborne units and the Soviets use the SD-44 85mm gun, a weapon that has a small auxiliary engine to drive it around the battlefield.

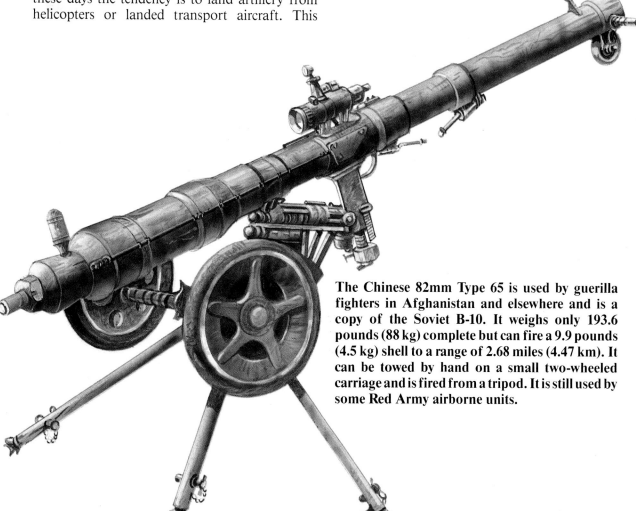

The Chinese 82mm Type 65 is used by guerilla fighters in Afghanistan and elsewhere and is a copy of the Soviet B-10. It weighs only 193.6 pounds (88 kg) complete but can fire a 9.9 pounds (4.5 kg) shell to a range of 2.68 miles (4.47 km). It can be towed by hand on a small two-wheeled carriage and is fired from a tripod. It is still used by some Red Army airborne units.

18 Mountain Guns

Airborne artillery is not the only type that is required to be light but strong, for mountain warfare is another form of combat that requires such specialized equipment. There are many parts of the world where fighting can be expected to take place in mountainous country — Norway is one example. In mountain warfare, troops may not always be expected to climb mountains to get into action but they will have to travel light over areas where there are few roads and only rough and steep tracks.

In such conditions, artillery is still required, and special types have been developed. These special weapons are known as pack artillery for they are designed to be broken down into a number of loads that are light enough for pack animals such as mules to carry; and at times when animals cannot be used, men will have to carry the guns. The design of pack artillery is specialized because it has to be possible to strip down the weapon, usually a howitzer, into small parts which are light enough to be carried, and then put the weapon back together again equally quickly. Once assembled it has to be safe to fire. When stripped down the various parts are carried by special harnesses on pack animals. When men have to do the carrying, some parts, such as barrels, are too large to be carried by one man so two have to share the burden between them.

A typical mountain artillery design still in use today is the Italian 105mm Model 56 Pack Howitzer. This versatile weapon was at one time used by the British Army as a field howitzer but for the mountain artillery role it can be broken down into eleven basic parts, the heaviest of which weighs 268.4 pounds (122 kg). Using special backpacks

The OTO Melara 105mm Model 56 pack howitzer has a cranked axle that allows it to be raised for normal firing (A) or lowered to reduce its height for the anti-tank role where concealment is important. In the lower position (B) barrel elevation is limited but traverse to "follow" tank targets is increased.

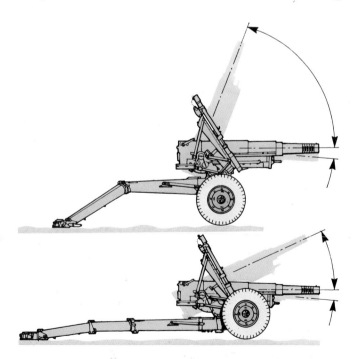

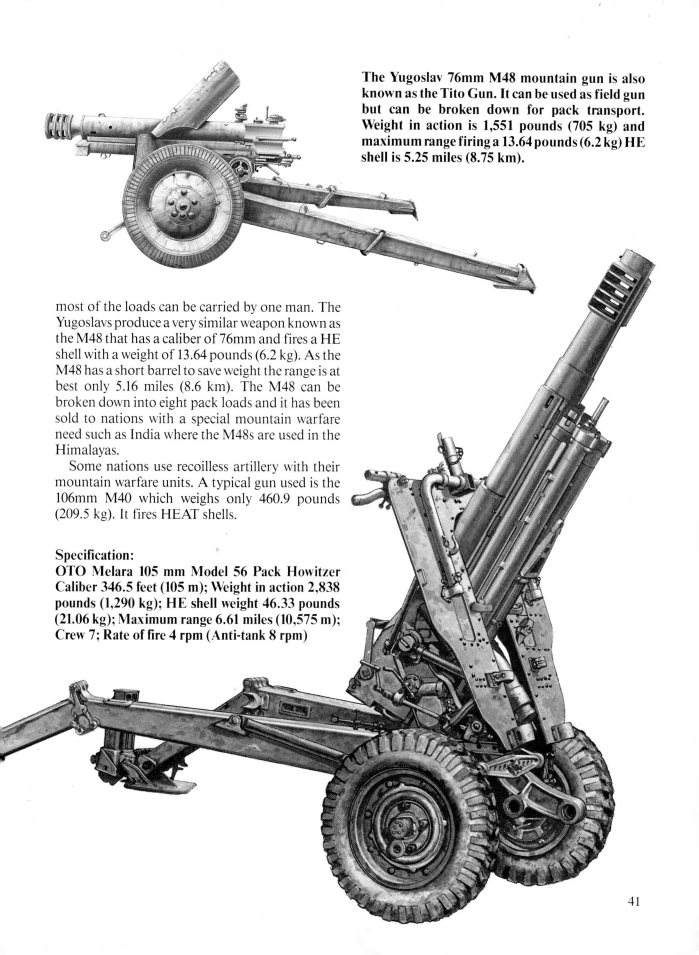

The Yugoslav 76mm M48 mountain gun is also known as the Tito Gun. It can be used as field gun but can be broken down for pack transport. Weight in action is 1,551 pounds (705 kg) and maximum range firing a 13.64 pounds (6.2 kg) HE shell is 5.25 miles (8.75 km).

most of the loads can be carried by one man. The Yugoslavs produce a very similar weapon known as the M48 that has a caliber of 76mm and fires a HE shell with a weight of 13.64 pounds (6.2 kg). As the M48 has a short barrel to save weight the range is at best only 5.16 miles (8.6 km). The M48 can be broken down into eight pack loads and it has been sold to nations with a special mountain warfare need such as India where the M48s are used in the Himalayas.

Some nations use recoilless artillery with their mountain warfare units. A typical gun used is the 106mm M40 which weighs only 460.9 pounds (209.5 kg). It fires HEAT shells.

Specification:
OTO Melara 105 mm Model 56 Pack Howitzer
Caliber 346.5 feet (105 m); Weight in action 2,838 pounds (1,290 kg); HE shell weight 46.33 pounds (21.06 kg); Maximum range 6.61 miles (10,575 m); Crew 7; Rate of fire 4 rpm (Anti-tank 8 rpm)

19 Fire Control and Radar

Modern artillery is so accurate that computers can be used to control their fire. With the aid of computers, gunners can now work out calculations that once had to be solved by looking up endless lists of tables and making use of slide rules. Today every battery section has its own fire-control computer, and on some self-propelled weapons every vehicle has a computer. Into these computers are entered all sorts of factors, such as the target range, its position relative to the battery, the height at which the weapons are being fired, the outside weather conditions such as wind strength and temperature, the types of ammunition to be fired, the slight aiming differences that always exist between weapons in the same battery, and so on. Within a few seconds all these variables can be used by the fire-control computer to produce accurate information with which to aim the weapons. On some fire-control systems the information is fed direct to the weapon, ready for the aimer to enter into his gun sights and lay the barrel. In other systems it is read off the computer by a gunner and sent to the guns by telephone or radio.

Using fire-control computers, target information for the weapons in a battery can be rapidly calculated, but the computer can only use the information that is supplied to it by various sources. The forward observer is one input to the computer, but radar and laser devices may also be employed. Radar can be used to discover the position of enemy artillery in order to direct counter-battery fire against it. Some of these radars are quite large and complex instruments. A typical example is the American AN/TPQ-36 locating radar that is set up near the front lines. It sweeps the enemy positions, trying to detect shells that are being fired. Once a shell in flight has been located, the radar can work out exactly where the shell has been fired from and send the information to a battery. Other radars are used to detect moving tank formations.

Lasers are used by forward observers to discover the exact ranges of targets, and sometimes they are mounted directly on weapons for this purpose.

A British Army Field Artillery Computer Equipment (FACE) installed in a Land Rover

A mortar-detecting radar in action. It projects a radar beam over suspected enemy positions and when a mortar bomb is detected by the beam it automatically projects a second beam to detect the bomb as it moves upwards. By using a computer the crew can then determine exactly where the enemy mortar is located and direct counter-battery fire.

Trajectory of mortar bomb

Time between intercepts measured

Beam elevated

Reflector radar

Alert position

Beam lowered

Radar horn

Enemy mortar

Radar equipment mounting

Main cable

Monitor indicator

Co-ordinate indicator

20
The Future

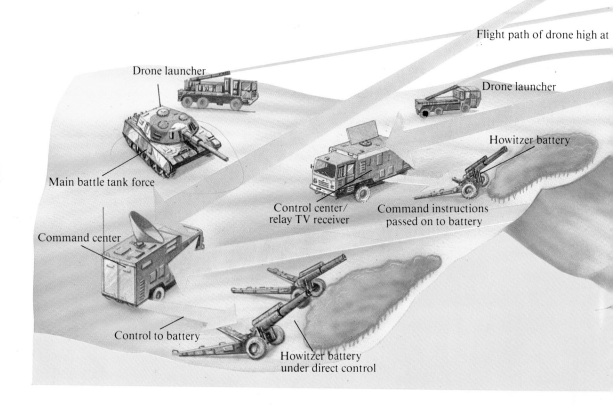

Flight path of drone high at

Drone launcher

Drone launcher

Howitzer battery

Main battle tank force

Control center/
relay TV receiver

Command instructions
passed on to battery

Command center

Control to battery

Howitzer battery
under direct control

Artillery is still developing and in the future it seems probable that it will become even more powerful than it is now. For most large armies the self-propelled weapon will probably replace the towed weapon except for special purposes such as airborne warfare. It can be predicted safely that ranges will increase from the 14.4-18 miles (24-30 km) of today, to well over 24 miles (40 km). This may be done by several means, including shell design and longer barrels, but perhaps the greatest advances will be produced by liquid propellants.

Current propellant charges are solid and are loaded into the weapon in bags or metal cases. When using liquid propellants, a shell will be loaded normally and the breech closed. A liquid

will then be injected into the space behind the shell and ignited. The resultant energy will be far more powerful and more sustained than that produced by solid propellants. This will ensure that the shell moves along the barrel at a faster rate and is propelled by a more constant pressure so it will travel further once it leaves the muzzle. The liquid propellant could be carried in a tank near the barrel and replaced in the same way fuel is poured into a vehicle tank, making supply and reloading much easier.

The future self-propelled weapon will be mounted on a much more mobile platform than today's tracked vehicles. Powered by gas turbines, the future self-propelled vehicle will be much

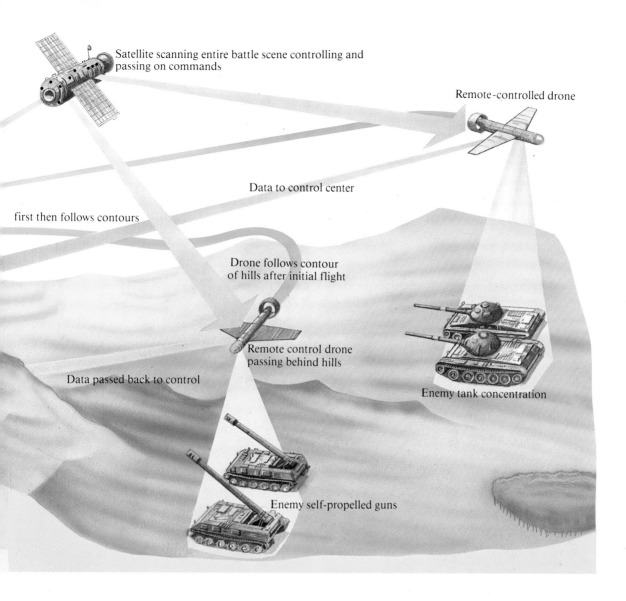

Satellite scanning entire battle scene controlling and passing on commands

Remote-controlled drone

Data to control center

first then follows contours

Drone follows contour of hills after initial flight

Remote control drone passing behind hills

Enemy tank concentration

Data passed back to control

Enemy self-propelled guns

smaller but have a higher output engine to provide more speed and better cross-country performance. The onboard fire-control systems will be more powerful as well, taking in information and using target information from space satellites hovering overhead. More target information will be available from pilotless drone aircraft buzzing over enemy positions.

Fewer men will be used to actually control and fire the weapons. The use of liquid propellants will make the use of automatic loaders much more likely, and each self-propelled weapon may have a crew of just two or three men. In the battery headquarters sophisticated computers will make more and more fire-control decisions and be used in-

creasingly to maintain communications and keep stocks of ammunition and supplies. But in the batteries, the guns and howitzers will still be there firing away at a distant enemy.

On the battlefield of the future artillery batteries will use computers, drones and even space satellites to either detect targets or direct communication links. Future artillery battles may be fought at ranges much greater than those used now.

Glossary

AP
Armor Piercing, a term applied to special solid projectiles intended to punch their way through tank or self-propelled gun armor

APDS
Armor Piercing Discarding Sabot, a special form of armor piercing projectile where the projectile is smaller than the caliber of the gun from which it is fired. The space between the projectile and the sides of the barrel is taken up by spacers called sabots and these fall away from the projectile once it has left the muzzle .

APDSFS
Armor Piercing Discarding Sabot Fin Stabilised, a special form of APDS on which fins are added at the tail of the projectile to stabilize it during flight. Usually used on tank guns but also used with some forms of artillery.

BASE BLEED
A special form of unit attached to the base of an artillery projectile to produce more range. In flight the base bleed unit burns to produce gases around the rear of the projectile and these gases prevent air currents forming around the tail of the projectile that would cause drag and thus reduce possible range.

BOMBLET
A small explosive device carried inside a cargo round and scattered from the body of the carrier projectile either in flight or when the projectile strikes the ground. Bomblets may be very small and may either detonate as they strike the ground or a target or lie around to act as small land mines.

BREECH
The part of a gun or howitzer into which the ammunition is placed before firing and which is then closed to seal off the rear of the chamber and barrel. The mechanism used for the breech is usually one of two main types, the sliding block in which a block of metal slides across (or up and down) to close the aperture through which the ammunition is loaded or of the screw type in which a piece of metal is screwed into the aperture.

CARGO ROUNDS
The term cargo round denotes a special form of projectile that contains some other form of payload besides explosive. A typical load could be a number of bomblet or land mines, smoke canisters or a canister containing flares.

CALIBER
The method of determining the size of the gun or howitzer barrel interior which is measured across the lands of the barrel

CHAMBER
The part of the gun or howitzer barrel closest to the breech in which the charge used to propel the projectile from the barrel is placed and detonated. It is smooth-bored and the metal surrounding it is always the thickest part of the barrel.

CRADLE
The mechanism of the gun in which the barrel is slung to allow it to be moved up and down and sometimes from side to side for aiming

DRONE
A small pilotless aircraft controlled from the ground which is used to carry television or other cameras to observe enemy positions and targets. Many are now known as RPVs or remotely piloted vehicles.

ELEVATION
The angle to which a gun or howitzer barrel is raised or lowered for firing. As a general rule guns have low angles of elevation while howitzers can have very high elevation angles.

FAASV
Field Artillery Ammunition Support Vehicle, a special form of tracked and armored vehicle that is used to carry ammunition to self-propelled artillery batteries in the field and pass the ammunition to them without any crew members having to leave the protection of their vehicles

GUN
An artillery weapon on which the projectile and charge are loaded in one piece, which is used to fire projectiles at high muzzle velocities in flat trajectories

GUN-HOWITZER
An artillery weapon on which separate-loading ammunition can be used (as in a howitzer) but which can also fire at high muzzle velocities

and with low trajectories. Most modern artillery weapons now fall into this category.

HE
High Explosive, the usual filling for an artillery shell. It produces its destructive effect in two ways, using the blast produced when the high explosive is detonated to cause damage and by hurling the steel fragments of the shell body produced when the explosion occurs. These fragments are sometimes known as shrapnel.

HEAP
A special form of armor-piercing projectile know as High Explosive Armor Piercing. It is a special hard-nosed shell that can punch a hole through armor. Once through the armor a small payload of high explosive detonates. Also known as APHE.

HESH
High Explosive Squash Head, a type of armor-defeating projectile in which the nose of the projectile squashes against a target and then detonates. The resultant shock breaks off fragments of armor from within the target know as "scabs" and these produce damage inside the target.

HOWITZER
A type of artillery on which the projectile and propelling charge are loaded separately, allowing the size of the charge to be varied to suit the range and the type of target being fired at. A howitzer can fire its projectiles at high angles of elevation to allow them to drop almost vertically onto the target.

ILLUMINATING
A type of shell that contains powerful flares for producing a bright light to allow gunners to see targets at night. The flare is usually ejected from the carrier shell while still in flight and the flare falls to the ground glowing and suspended from a small parachute.

LANDS
The raised areas between the rifling in a gun or howitzer barrel. The weapon's caliber is measured between opposing lands on either side of the barrel.

MORTAR
A modern mortar is described as a smooth-bored weapon that fires finned bombs at very high angles of elevation. Only the very largest caliber mortars come into the artillery category. At one time in the past a mortar was a special form of artillery with a very short barrel that fired shells at high angles of elevation but such weapons are now no longer in use.

MET
A term used to describe any dealing with meteorology in the artillery sense. Met factors such as wind and temperature affect the trajectory of an artillery projectile in flight and so have to be taken into account.

MUZZLE
The face of the barrel at the end from which the projectile emerges

MUZZLE BRAKE
A device fitted onto a muzzle to direct some of the gases produced on firing away to one side. This can have the effect of reducing recoil forces slightly.

MUZZLE VELOCITY
The speed at which a projectile leaves the muzzle, sometimes written as V_0, or velocity zero, or perhaps MV

OBTURATION
The method of sealing the breech at the instant of firing. Most artillery weapons rely on a breech block, a solid piece of metal at the bottom of the chamber that fits so well that no gas can escape to the rear. Other types of weapon may use a pad that expands under pressure to seal the edges of the breech.

PACK
A term used to indicate that the weapon involved can be broken down into loads for transport, as in pack howitzer

PROJECTILE
A term used to denote anything fired from a gun or howitzer, that is, the gunner's weapon

QUICK-FIRING
An old term used to indicate that the ammunition (ie projectile and propelling charge) was loaded in one piece. Used only in connection with guns, it is now a term that is obsolete.

RECOILLESS
A special type of gun in which the gases produced on firing are allowed to escape to the rear in such a manner that no recoil forces are produced

RAP
Rocket Assisted Projectile, a type of projectile used to increase range by cutting in a rocket motor in the rear of the projectile just as it is losing forward speed at the high point of its trajectory. The resultant thrust from the rocket pushes the projectile further.

RECUPERATOR
The part of the gun or howitzer mechanism that pushes the barrel back to its original position after recoil forces have pushed it to the rear. Most recuperators use springs combined with some type of thick fluid in an enclosed tube.

RIFLING
The grooves inside a barrel that provide twist to a projectile to keep it stable in flight

ROUND
A general term that might be used to indicate either a projectile only or a one-piece combination of projectile and propelling charge. It may also be used to describe the number of times a gun has been fired.

RPV
Remotely Piloted Vehicle, another term for drone

SHELL
A term used to describe a projectile with a hollow interior into which high explosive or some other payload can be packed

SHOT
A solid artillery projectile that is usually used for armor piercing

SMOKE
A special form of projectile that contains a payload that produces smoke when detonated. The smoke may be used to either produce a smoke screen or colored smoke may be used to indicate a target to other gunners or to strike aircraft.

SP
Self Propelled

TRAJECTORY
The path an artillery projectile follows when in flight. Gun trajectories are usually flat while howitzer trajectories can be high and curved.

TRAVERSE
The amount a gun barrel can be aimed to either side of a central line

Index